VINTON ELEMENTARY SCHOOL-MANSFIELD
3 4057 13033 9450

Y0-BZI-315

DATE DUE			

Natural Resources

MINERALS

Emma Bassier

DiscoverRoo
An Imprint of Pop!
popbooksonline.com

abdobooks.com

Published by Pop!, a division of ABDO, PO Box 398166, Minneapolis, Minnesota 55439.

Printed in the United States of America, North Mankato, Minnesota.

102019
012020

THIS BOOK CONTAINS RECYCLED MATERIALS

Cover Photo: iStockphoto
Interior Photos: iStockphoto, 1, 6, 7, 8, 9 (diamond), 9 (corundum), 9 (topaz), 9 (quartz), 9 (orthoclase), 9 (apatite), 9 (fluorite), 9 (calcite), 9 (gypsum), 9 (talc), 11, 13, 14, 15, 16–17, 18 (top), 18 (bottom), 19 (top), 19 (bottom), 23, 24, 27, 31; Shutterstock Images, 5, 10, 22, 25, 28, 30; Carla Gottgens/Bloomberg/Getty Images, 21; Monty Rakusen/Cultura Creative (RF)/Alamy, 29

Editor: Sophie Geister-Jones
Series Designer: Jake Slavik

Library of Congress Control Number: 201994248

Publisher's Cataloging-in-Publication Data

Names: Bassier, Emma, author.

Title: Minerals / by Emma Bassier

Description: Minneapolis, Minnesota : Pop!, 2020 | Series: Natural resources | Includes online resources and index.

Identifiers: ISBN 9781532165856 (lib. bdg.) | ISBN 9781532167171 (ebook)

Subjects: LCSH: Minerals--Juvenile literature. | Natural resources--Juvenile literature. | Environment--Juvenile literature. | Ecology--Juvenile literature. | Geology--Juvenile literature.

Classification: DDC 333.85--dc23

WELCOME TO DiscoverRoo!

Pop open this book and you'll find QR codes loaded with information, so you can learn even more!

Scan this code* and others like it while you read, or visit the website below to make this book pop!

popbooksonline.com/minerals

*Scanning QR codes requires a web-enabled smart device with a QR code reader app and a camera.

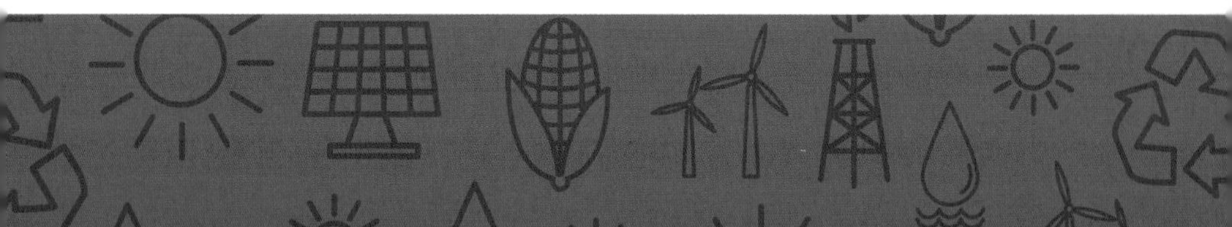

TABLE OF
CONTENTS

CHAPTER 1

WHAT IS A MINERAL?

A boy writes with a pencil. Its tip is made from graphite. A woman puts on a gold necklace. A girl shakes salt on her food. Graphite, gold, and salt are all minerals.

WATCH A VIDEO HERE!

Water often contains the mineral salt. When the water dries up, salt is left behind.

Minerals are solid substances. They occur naturally on Earth.

Many minerals are found in rocks in the ground. Some rocks, such as limestone, are made of only one mineral. Other rocks are made of multiple minerals. For example, granite contains the minerals quartz, feldspar, and mica.

Limestone is a rock made from a mineral called calcite.

Gems, such as rubies, are rare minerals that people sometimes polish.

Muscovite is a mineral that breaks into sheets.

More than 5,000 different minerals exist on Earth. Each has its own **properties**. These properties include color, hardness, and cleavage. Cleavage describes how a mineral breaks. Some minerals break into flat sheets. Others break randomly into odd-shaped pieces.

MINERAL HARDNESS

The Mohs hardness scale measures how hard a mineral is. It uses numbers from 1 to 10. Minerals with higher numbers are harder to scratch.

HARDEST

10 MOHS - DIAMOND

9 MOHS - CORUNDUM

8 MOHS - TOPAZ

7 MOHS - QUARTZ

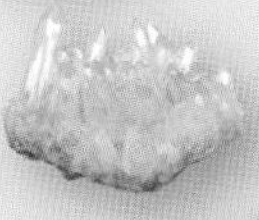

6 MOHS - ORTHOCLASE

5 MOHS - APATITE

4 MOHS - FLUORITE

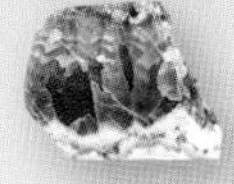

3 MOHS - CALCITE

2 MOHS - GYPSUM

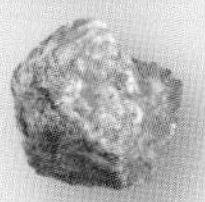

1 MOHS - TALC

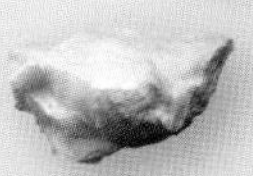

SOFTEST

A mineral's properties decide how it can be used. For example, talc is a soft mineral. It is easily pressed into a powder. It can soak up oil and smells. For these reasons, talc is often used as foot powder.

Powder made from talc can soak up liquid.

Diamonds are extremely hard minerals. Cutting them requires special saws.

DID YOU KNOW?

Diamonds are a mineral. They are the hardest substance found naturally on Earth.

CHAPTER 2

MINERALS EVERYWHERE

Minerals make up objects people use in daily life. They make up parts of cups and toothbrushes. Many electronic devices need parts made from minerals to work. Minerals are also in some foods.

LEARN MORE HERE!

Batteries often contain minerals.

DID YOU KNOW?

The mineral silver slows down bacteria growth. It is often added to medicine such as penicillin.

Panning in rivers and streams helps people separate gold from other rocks.

People **pan** for some minerals. But most minerals come from mines. People dig into rock to get minerals from it. First, they use large machines to drill

into the ground or water. Then they bring ore up to the surface. Ore is rock that contains a form of metal or mineral.

TYPES OF MINING

Surface, underground, and marine are three types of mining. Surface mining is the most common type of mining. People dig large, open pits into the earth. In underground mining, people drill deep tunnels called shafts. Marine mining is the least common. People mostly use this method to dig salts from the beach or seafloor.

After ore comes out of the ground, the rock and mineral parts must be separated. The process often happens at buildings called mills. There, the ore

The process of separating minerals often requires large machines.

is cleaned, crushed, and heated. The process changes slightly depending on the mineral. Sometimes water or chemicals are added.

MINERALS THROUGH THE YEARS

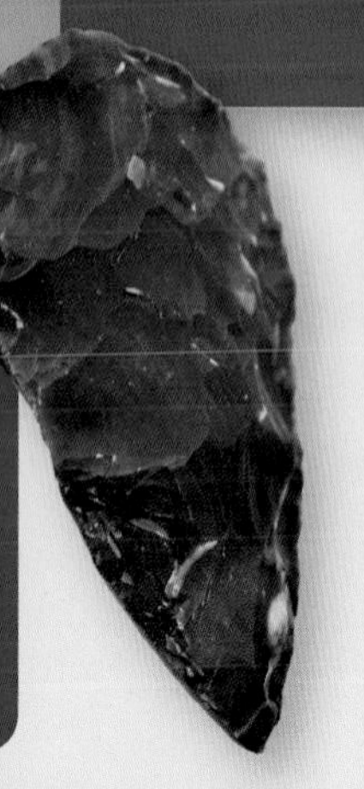

8000 BCE

People begin using flint to make tools such as scrapers and knives.

1766–256 BCE

People in China first use minerals in medicine and when burying the dead. They believe minerals have healing powers.

2500 BCE

Ancient Egyptians build pyramids using minerals such as limestone, sandstone, and granite.

1800s

People use drills and dynamite to blow up rocks. This makes mining minerals easier.

1556 CE

German scientist Georgius Agricola writes *De re metallica*. The book describes methods for mining minerals.

1900s

Scientists find new ways to separate minerals. They develop rare-earth mining. Rare-earth minerals are used to make parts of mobile phones, cameras, rockets, and more.

CHAPTER 3

RUNNING OUT

Many minerals exist in only a few places on Earth. These locations can be hard to reach. Mining can be difficult and dangerous. It can also be expensive. It may require many machines and workers.

COMPLETE AN ACTIVITY HERE!

People wear hard hats for protection when entering a mine.

Mining salt requires a lot of hard work. Salt is one mineral that humans need for survival.

In addition, minerals are **nonrenewable**. Earth has a limited supply of each one. Too much mining can cause minerals to run out. It can also damage the environment.

DID YOU KNOW?

Many **rare-earth** minerals are found in China. The country mines 95 percent of the supply used by people around the world.

Some people want to mine asteroids for rare minerals.

Mining often requires clearing land. People change the landscape when they dig and drill. Plants and animals may struggle to survive. Mines also create waste that harms plants and other living

Mining requires a lot of water. It can cause shortages and pollute nearby waterways.

Mines change the landscape. Sometimes, the changes can cause increased erosion.

creatures. The waste causes **pollution**. It can make people or animals sick. It can also kill plants.

CHAPTER 4

CONSERVING MINERALS

People are working to **conserve** minerals. Materials such as iron and steel are made from minerals. Instead of mining more, people reuse minerals that have already been mined.

LEARN MORE HERE!

Many electronic devices are made with rare-earth minerals. Scientists are searching for ways to recycle them.

Metal objects are often made from recycled materials.

People are trying to find new ways to make products. Mines are looking at better ways to get the minerals from the ground. They do not want to create as much waste. All over the world, scientists are looking for ways to not rely as much on minerals.

Many underground mines are large enough for machines to drive through.

People can recycle scrap metal, such as steel, as an alternative to mining more ore.

DID YOU KNOW?

Car companies in Japan recycle minerals from old batteries and electronics.

MAKING CONNECTIONS

TEXT-TO-SELF

There are many different kinds of minerals. Which minerals have you used or seen?

TEXT-TO-TEXT

Have you read other books about natural resources? How were those resources different from minerals?

TEXT-TO-WORLD

Earth has a limited supply of minerals. What is one thing you wouldn't be able to do if minerals ran out?

GLOSSARY

conserve – to save and not waste.

nonrenewable – having a limited amount.

pan – to separate minerals from rocks or soil by rinsing with water.

pollution – harmful substances that collect in the air, water, or soil.

property – a physical trait.

rare-earth – a type of uncommon mineral that is found deep in the earth and is difficult to mine.

INDEX

Scan this code* and others like it while you read, or visit the website below to make this book pop!

popbooksonline.com/minerals

*Scanning QR codes requires a web-enabled smart device with a QR code reader app and a camera.